Fifth Grade Math Quiz

By Greg Sherman

Home School Brew Press

www.HomeSchoolBrew.com

© 2013. All Rights Reserved.

Cover Image © jogyx - Fotolia.com

Table of Contents

About Us 5
Algebra 6
   Quiz 7
Data Analysis 8
   Quiz 9
Decimals 12
   Quiz 13
Fractions 14
   Quiz 15
Geometry 16
   Quiz 17
Measurement 19
   Quiz 20
Multiplying Large Numbers 22
   Quiz 23
Probability 24
   Quiz 25
Whole Numbers 26
   Quiz 27
Word Problems 28
   Quiz 29
Answer Key 30
   Algebra 31
      Quiz 32
   Data Analysis 33
      Quiz 34
   Decimals 36
      Quiz 37
   Fractions 38
      Quiz 38
   Geometry 39
      Quiz 40
   Measurement 42

Quiz	43
Multiplying Large Numbers	45
Quiz	45
Probability	46
Quiz	47
Whole Numbers	49
Quiz	50
Word Problems	51
Quiz	52

Disclaimer

This book was developed for parents and students of no particular state; while it is based on common core standards, it is always best to check with your state board to see what will be included on testing.

About Us

Homeschool Brew was started for one simple reason: to make affordable Homeschooling books! When we began looking into homeschooling our own children, we were astonished at the cost of curriculum. Nobody ever said homeschool was easy, but we didn't know that the cost to get materials would leave us broke.

We began partnering with educators and parents to start producing the same kind of quality content that you expect in expensive books...but at a price anyone can afford.

We are still in our infancy stages, but we will be adding more books every month. We value your feedback, so if you have any comments about what you like or how we can do better, then please let us know!

To add your name to our mailing list, go here: http://www.homeschoolbrew.com/mailing-list.html

Algebra

Quiz

1. Identify the coefficient in the equation $4t - 3 = 5$ _____

2. There are 19 fruit trees in Kyle's backyard. 9 of them are apple trees and the rest are peach trees. Which equation represents the situation?

    $9x = 19$ $9 - x = 19$ $x + 9 = 19$

3. Solve for the variable: $x - 16 = 23$ _____

4. What is the variable in the equation $17 + r = 74$? _____

5. Solve for the variable: $3x + 12 = 36$ _____

6. Identify the coefficient in the equation $20x = 200$ _____

7. Solve for the variable: $5y - 6 = 14$ _____

8. There are 8 kittens in a litter. 3 are black and the rest are white. Which equation represents the situation?

    $3 - x = 8$ $3 + x = 8$ $3x = 8$

9. Solve for the variable: $4x = 16$ _____

10. Solve for the variable: $2x + 9 = 29$ _____

Data Analysis

Quiz

Daphne is collecting data on the time it takes for her to do homework each night. She has her information in a table.

Night	Time to do Homework
1	22 minutes
2	27 minutes
3	35 minutes
4	27 minutes
5	22 minutes
6	27 minutes
7	16 minutes

1. What is the median for Daphne's data? _____

2. What is the range for Daphne's data? _____

3. What is the mode for Daphne's data? _____

4. What is the mean for Daphne's data? _____

5. Should Alyssa use measurement or observation to find the length of driveways in her neighborhood? _____

6. Tracy keeps track of the number of birds who eat at her bird feeder every day for a week. (4, 12, 6, 8, 4, 7, 9) What is the average number of birds who eat at her bird feeder? _____

7. Crista loves to run. She keeps track of how many miles she runs over five days. (15, 8, 9, 12, 7) What is the range? _____

8. Given the following set of data, what is the mode? (7, 12, 6, 12, 9, 5, 8, 5)
   _____

9. Eddie reads every night before he goes to bed and keeps track of the number of pages he reads. (34, 29, 12, 18, 25) What is the median? _____

10. Create a graph using the following table:

Day	Minutes Practiced the Violin
1	25 minutes
2	20 minutes
3	30 minutes
4	35 minutes
5	40 minutes
6	35 minutes

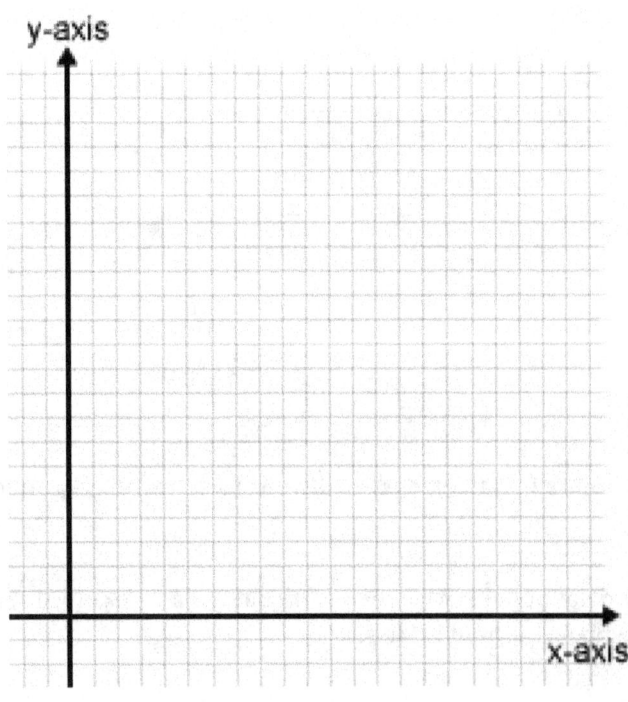

Decimals

Quiz

1. Compare the two numbers: 34.67   ⓧ 3.467

2. Which number is larger?   587.23   or   587.32   _____

3. What is the value of the number 3 in 901.234?   _____

4. Add 219.08 and 76.341   _____

5. Put the numbers in descending order: 98.342, 98.38, 97.423, 97.03
   _____

6. Compare the two numbers: 0.213   ⓧ 0.22

7. 56.7 – 52.82 =   _____

8. Which number is smaller?   4.623   or   .4623   _____

9. Write the decimal .741 as a fraction   _____

10. What is the value of the number 4 in 564.87?   _____

Fractions

Quiz

1. Which fraction is equivalent to $9/15$? $18/30$ $15/22$ $18/20$

2. Reduce $21/4$ _____

3. Which fraction is larger? $3/6$ or $7/12$ _____

4. $6/12 + 5/6 =$ _____

5. Find a fraction that is equivalent to $2\ 3/5$ _____

6. $15/4 \times 2/9 =$ _____

7. Reduce $29/6$ _____

8. $5/9 - 2/18 =$ _____

9. Which fraction is equivalent to $6/18$? $7/9$ $2/6$ $3/6$

10. $10/12 \div 4/9 =$ _____

Geometry

Quiz

1. What is the measure of the angle? _____

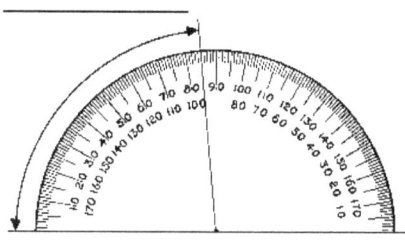

2. What type of angle is this?    _____

3. What type of angle is exactly 180°? _____

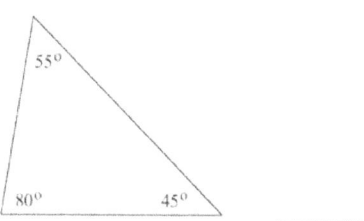

4. What type of triangle is this?   _____

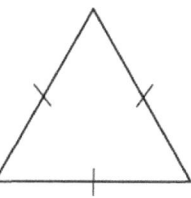

5. Is this triangle scalene or equilateral?   _____

6. An isosceles triangle has _____ equal sides and _____ equal angles.

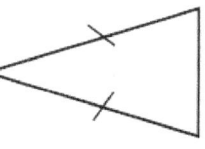

7. What type of triangle is this?   _____

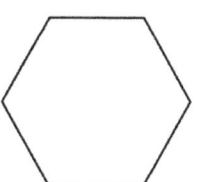

8. What are the critical attributes of this hexagon?   _____

9. Which prism is a hexagonal pyramid? 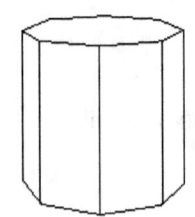 or

10. Which shape is not three-dimensional?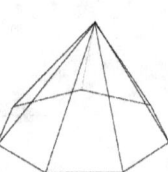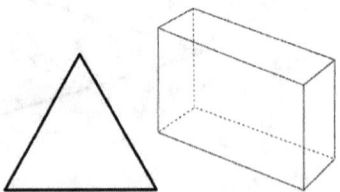

Measurement

Quiz

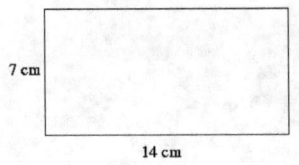

1. Figure out the area of the rectangle. _____

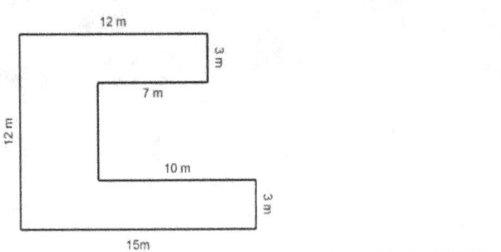

2. What is the perimeter of the polygon? _____

3. Convert 1.2 liters to milliliters. _____

4. How many centimeters are in 3.46 meters? _____

5. What temperature is displayed on the thermometer in Fahrenheit?

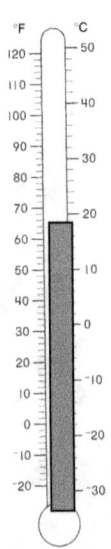

_____

6. How many hours pass between 3:15PM and 2:30AM? _____

7. How many kilometers are there in 5600 meters? _____

8. Find the volume of the prism. 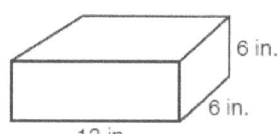 _____

9. Convert 21 feet to yards. _____

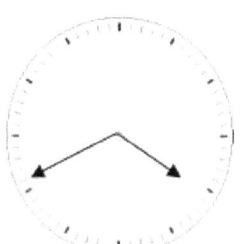

10. What time is it? _____

Multiplying Large Numbers

Quiz

1. 583
    × 7
    ----

2. 719
    × 4
    ----

3. 2013
    × 3
    ----

4. 827
    × 15
    ----

5. 4099
    × 18
    ----

6. 1753
    × 24
    ----

7. 2591
    × 347
    ----

8. 1277
    × 233
    ----

9. 874
    × 503
    ----

10. 6183
    ×782

Probability

Quiz

1. If there are 12 colors on a color wheel, what is the probability of landing on each individual color? _____

2. What is the probability, in a percentage, of 6/14? _____

3. If there is a 4/8 chance of something happening, what is the likelihood that it will happen? _____

4. There is a 1/6 chance that you will roll a 1, 2, 3, 4, 5, or 6 when rolling a dice. What is the probability that you will roll a number less than 4? _____

5. Place the following probability outcomes on the probability line: 0/7, 4/7, 6/7

    ```
    0           1/2          1
    |------------|------------|
    ```

6. Mark has different uniforms for his job. He has 7 white shirts, 9 blue shirts and 12 striped shirts. List all of the possible outcomes for the type of shirt that Mark could get when he reaches in his uniform drawer.

    _____

    _____

    _____

7. If there is a 85% chance of rain, what is the likelihood that it will rain? _____

8. There are 25 marbles in a jar with an equal number of orange, blue, white, green and yellow marbles. What is the probability of getting an orange marble? _____

9. What is the probability, in a percentage, of 6/18? _____

10. There is a 1/6 chance of getting a vanilla, chocolate, twist, raspberry, peanut butter or blackberry ice cream cone. What is the probability that you will choose a berry flavored ice cream cone? _____

Whole Numbers

Quiz

1. Is 15 a prime or composite number? _____

2. Round the answer to 97 ÷ 5   _____

3. What are the factors pairs for 32? _____

4. What is the value of 2 in the number 298,539? _____

5. Estimate the answer to 34 + 15 + 89   _____

6. Write the number 989 in expanded form   _____

7. Round the answer to 832 – 513 _____

8. Is 31 a prime or composite number?   _____

9. What are the factor pairs for 42? _____

10. Write the number 781 in written form   _____

Word Problems

Quiz

1. Ilene is an artist. She makes ceramic bowls and sells them at craft fairs. Ilene has 27 bowls to bring with her to the craft fair on Saturday. At the craft fair, Ilene sells all of her bowls. Ilene also gets orders for 19 bowls. It takes Ilene 3 hours to make each bowl. How long will it take her to make the bowls that were ordered? _____

2. Henry is helping at a spaghetti dinner. He is in charge of making the juice. Henry has to use 13 gallons of water to make the juice every half an hour. After 3 hours, estimate how many gallons of water will Henry have used? _____

3. Pam loves to paint. She has 56 different color paints. Pam wants to put her paints in containers. The containers only hold 8 colors each. Write an equation and then solve to figure out how many containers will Pam need? _____

4. You are learning to scuba dive. You start in waters that are 62.6° As you start to dive down the water gets colder. It is 11.7° colder where you dive than it is at the surface. What is the temperature of the water you are diving in? _____

5. You are timing a race. The first person passes the finish line in 23 minutes and 16 seconds. The last person passes the finish line in 37 minutes and 29 seconds. How much faster was the first person than the last person? _____

6. Your teacher has 5 different types of stickers that she puts on completed homework. She has 15 smiley faces, 20 flowers, 20 hearts, 15 stars and 25 animals. What is the probability that you will get a heart sticker on your paper? _____

7. There are 58 photos printed. You have to print 3 times more. Write an equation and then solve to figure out how many photos total you have to print. _____

8. Kyle is planning a senior prank. He is going to cover the football field with spray paint. He knows the area of a field is 121 meters^2. The sides of the field are the same. Kyle needs to know what the lengths of the sides of the field are so that he can make sure he has enough people on each side to get the field completely covered. _____

9. Kristie is making cookies for a surprise party. There will be a lot of people there so Kristie needs to triple the recipe. The original recipe calls for ½ tsp. baking soda. How much baking soda does Kristie need now? _____

10. I am three-dimensional. I have 6 faces and 6 pairs of parallel lines. Each face has right angles. All of my faces are congruent with each other. Who am I? _____

Answer Key

Algebra

Quiz

11. Identify the coefficient in the equation 4t − 3 = 5   ____4____

12. There are 19 fruit trees in Kyle's backyard. 9 of them are apple trees and the rest are peach trees. Which equation represents the situation?

    $9x = 19$ $9 - x = 19$ **(x + 9 = 19)**

13. Solve for the variable: x − 16 = 23   ____7____

14. What is the variable in the equation 17 + r = 74?   ____r____

15. Solve for the variable: 3x + 12 = 36   ____8____

16. Identify the coefficient in the equation 20x = 200   ____20____

17. Solve for the variable: 5y − 6 = 14   ____4____

18. There are 8 kittens in a litter. 3 are black and the rest are white. Which equation represents the situation?

    $3 - x = 8$ **(3 + x = 8)** $3x = 8$

19. Solve for the variable: 4x = 16   ____4____

20. Solve for the variable: 2x + 9 = 29   ____10____

Data Analysis

Quiz

Daphne is collecting data on the time it takes for her to do homework each night. She has her information in a table.

Night	Time to do Homework
1	22 minutes
2	27 minutes
3	35 minutes
4	27 minutes
5	22 minutes
6	27 minutes
7	16 minutes

11. What is the median for Daphne's data?   _____27_____

12. What is the range for Daphne's data?   _____19_____

13. What is the mode for Daphne's data?   _____27_____

14. What is the mean for Daphne's data?   _____25_____

15. Should Alyssa use measurement or observation to find the length of driveways in her neighborhood?
    ___measurement___

16. Tracy keeps track of the number of birds who eat at her bird feeder every day for a week. (4, 12, 6, 8, 4, 7, 9) What is the average number of birds who eat at her bird feeder?
    _____7_____

17. Crista loves to run. She keeps track of how many miles she runs over five days. (15, 8, 9, 12, 7) What is the range?   _____8_____

18. Given the following set of data, what is the mode? (7, 12, 6, 12, 9, 5, 8, 5)
    ___12 and 5___

19. Eddie reads every night before he goes to bed and keeps track of the number of pages he reads. (34, 29, 12, 18, 25) What is the median? _____25_____

20. Create a graph using the following table:

Day	Minutes Practiced the Violin
1	25 minutes
2	20 minutes
3	30 minutes
4	35 minutes
5	40 minutes
6	35 minutes

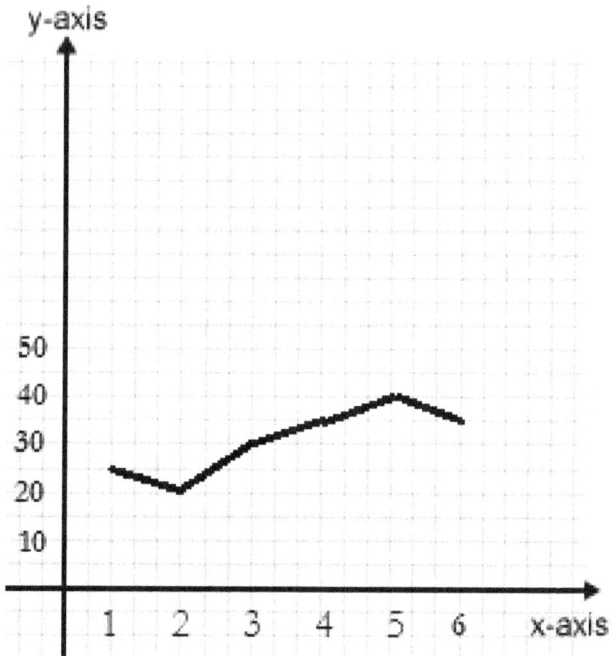

Decimals

Quiz

11. Compare the two numbers: 34.67 > **3.467**

12. Which number is larger? 587.23 or 587.32   587.32

13. What is the value of the number 3 in 901.234?   hundredths

14. Add 219.08 and 76.341   295.421

15. Put the numbers in descending order: 98.342, 98.38, 97.423, 97.03
    98.38, 98.342, 97.423, 97.03

16. Compare the two numbers: 0.213 < **0.22**

17. 56.7 − 52.82 =   3.88

18. Which number is smaller? 4.623 or .4623   .4623

19. Write the decimal .741 as a fraction   741/1000

20. What is the value of the number 4 in 564.87?   one

Fractions

Quiz

11. Which fraction is equivalent to $^9/_{15}$? $\underline{^{18}/_{30}}$ $^{15}/_{22}$ $^{18}/_{20}$

12. Reduce $^{21}/_4$ $\underline{5\,^1/_4}$

13. Which fraction is larger? $^3/_6$ or $\underline{^7/_{12}}$ _____

14. $^6/_{12} + ^5/_6 =$ $\underline{^{16}/_{12}}$

15. Find a fraction that is equivalent to $2\,^3/_5$ $\underline{^{13}/_5}$

16. $^{15}/_4 \times ^2/_9 =$ $\underline{^5/_6}$

17. Reduce $^{29}/_6$ $\underline{4\,^5/_6}$

18. $^5/_9 - ^2/_{18} =$ $\underline{^8/_{18}}$

19. Which fraction is equivalent to $^6/_{18}$? $^7/_9$ $\underline{^2/_6}$ $^3/_6$

20. $^{10}/_{12} \div ^4/_9 =$ $\underline{^{90}/_{36}}$

Geometry

Quiz

11. What is the measure of the angle? _____85°_____

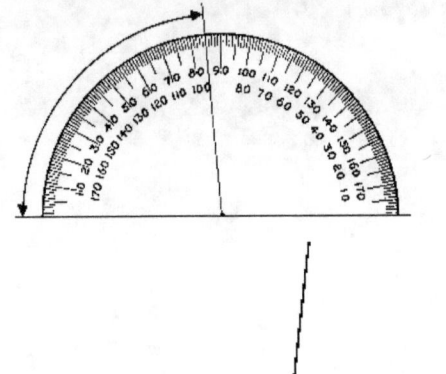

12. What type of angle is this? 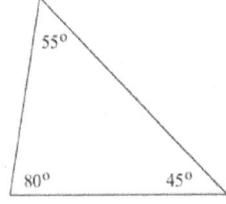 _____acute_____

13. What type of angle is exactly 180°? _____straight_____

14. What type of triangle is this? 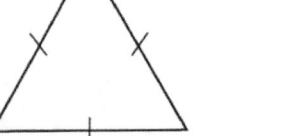 _____scalene_____

15. Is this triangle scalene or equilateral? 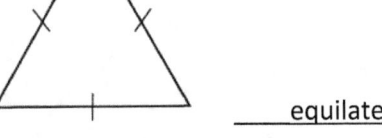 _____equilateral_____

16. An isosceles triangle has _____2_____ equal sides and _____2_____ equal angles.

17. What type of triangle is this?  _____isosceles_____

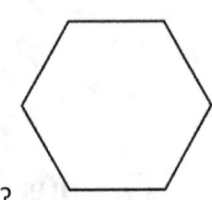

18. What are the critical attributes of this hexagon? _____6 sides, 6 acute angles and two-dimensional_____

19. Which prism is a hexagonal pyramid? 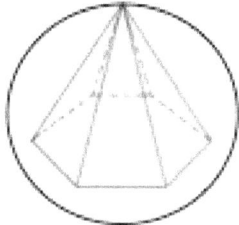 or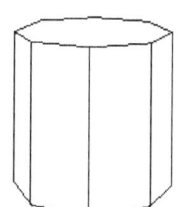

20. Which shape is not three-dimensional?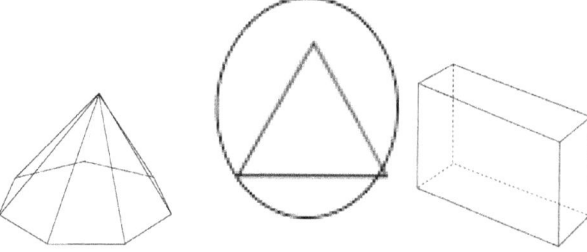

Measurement

Quiz

11. Figure out the area of the rectangle. 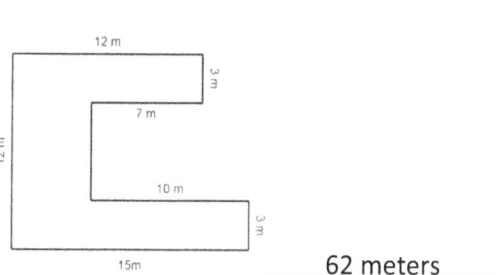   __98 cm²__

12. What is the perimeter of the polygon?   __62 meters__

13. Convert 1.2 liters to milliliters.   __1200__

14. How many centimeters are in 3.46 meters?   __346__

15. What temperature is displayed on the thermometer in Fahrenheit?

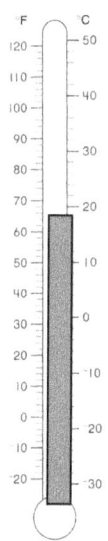

__66°__

16. How many hours pass between 3:15PM and 2:30AM?   __6 ¼ hours__

17. How many kilometers are there in 5600 meters?   __5.6__

18. Find the volume of the prism. 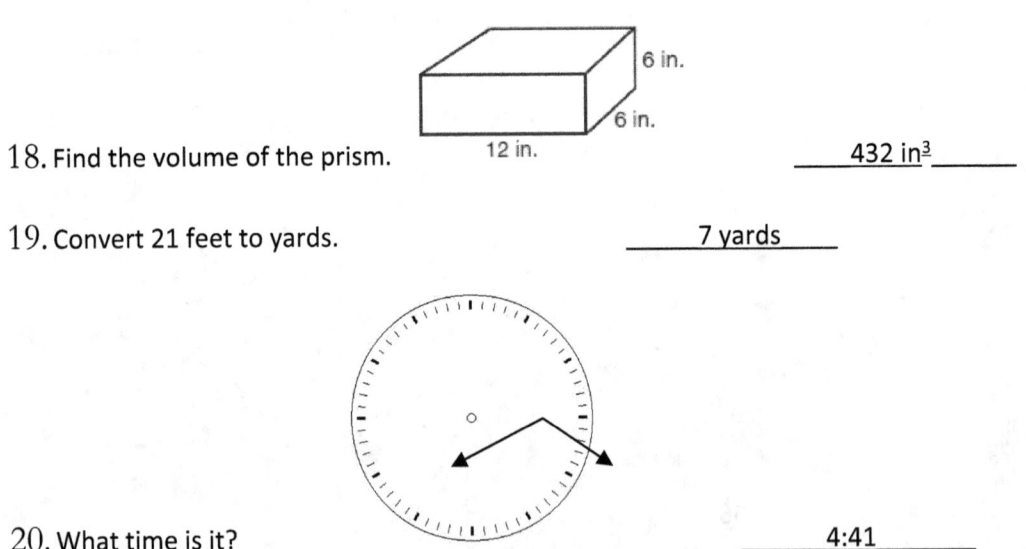   __432 in³__

19. Convert 21 feet to yards.   __7 yards__

20. What time is it?   __4:41__

Multiplying Large Numbers

Quiz

1. 583
    × 7

    4081

2. 719
    × 4

    2876

3. 2013
    × 3

    6039

4. 827
    × 15

    12405

5. 4099
    × 18

    73782

6. 1753
    × 24

    42072

7. 2591
    × 347

    899077

8. 1277
    × 233

    297541

9. 874
    × 503

    439622

10. 6183
    × 782

    4835106

Probability

Quiz

11. If there are 12 colors on a color wheel, what is the probability of landing on each individual color?
    <u>   1/12   </u>

12. What is the probability, in a percentage, of 6/14?   <u>   43%   </u>

13. If there is a 4/8 chance of something happening, what is the likelihood that it will happen?
    <u>even chance</u>

14. There is a 1/6 chance that you will roll a 1, 2, 3, 4, 5, or 6 when rolling a dice. What is the probability that you will roll a number less than 4? <u> 1/2 </u>

15. Place the following probability outcomes on the probability line: 0/7, 4/7, 6/7

    ```
    0              1/2              1
    |  0/7          |  4/7     6/7  |
    ```

16. Mark has different uniforms for his job. He has 7 white shirts, 9 blue shirts and 12 striped shirts. List all of the possible outcomes for the type of shirt that Mark could get when he reaches in his uniform drawer.

    <u>   white- 7/28   </u>

    <u>   blue- 9/28   </u>

    <u>   striped- 12/28   </u>

17. If there is a 85% chance of rain, what is the likelihood that it will rain?
    <u>   likely   </u>

18. There are 25 marbles in a jar with an equal number of orange, blue, white, green and yellow marbles. What is the probability of getting an orange marble?
    <u>   1/5   </u>

19. What is the probability, in a percentage, of 6/18?   <u>   33%   </u>

20. There is a 1/6 chance of getting a vanilla, chocolate, twist, raspberry, peanut butter or blackberry ice cream cone. What is the probability that you will choose a berry flavored ice cream cone?
    <u>   2/6   </u>

Whole Numbers

Quiz

11. Is 15 a prime or composite number?   _composite_

12. Round the answer to 97 ÷ 5   __20__

13. What are the factors pairs for 32?   _1 and 32, 2 and 16, 4 and 8_

14. What is the value of 2 in the number 298,539?   _hundred thousand_

15. Estimate the answer to 34 + 15 + 89   __140__

16. Write the number 989 in expanded form   _900 + 80 + 9_

17. Round the answer to 832 – 513   __300__

18. Is 31 a prime or composite number?   __prime__

19. What are the factor pairs for 42?   _1 and 42, 2 and 21, 3 and 14, 6 and 7_

20. Write the number 781 in written form   _seven hundred eighty one_

Word Problems

Quiz

11. Ilene is an artist. She makes ceramic bowls and sells them at craft fairs. Ilene has 27 bowls to bring with her to the craft fair on Saturday. At the craft fair, Ilene sells all of her bowls. Ilene also gets orders for 19 bowls. It takes Ilene 3 hours to make each bowl. How long will it take her to make the bowls that were ordered?   __57 hours__

12. Henry is helping at a spaghetti dinner. He is in charge of making the juice. Henry has to use 13 gallons of water to make the juice every half an hour. After 3 hours, estimate how many gallons of water will Henry have used?
         __30 gallons__

13. Pam loves to paint. She has 56 different color paints. Pam wants to put her paints in containers. The containers only hold 8 colors each. Write an equation and then solve to figure out how many containers will Pam need?
         __x = 7__

14. You are learning to scuba dive. You start in waters that are 62.6° As you start to dive down the water gets colder. It is 11.7° colder where you dive than it is at the surface. What is the temperature of the water you are diving in?
         __50.9 degrees__

15. You are timing a race. The first person passes the finish line in 23 minutes and 16 seconds. The last person passes the finish line in 37 minutes and 29 seconds. How much faster was the first person than the last person?
         __14 minutes and 13 seconds__

16. Your teacher has 5 different types of stickers that she puts on completed homework. She has 15 smiley faces, 20 flowers, 20 hearts, 15 stars and 25 animals. What is the probability that you will get a heart sticker on your paper?
         __20/95__

17. There are 58 photos printed. You have to print 3 times more. Write an equation and then solve to figure out how many photos total you have to print.
         __x = 174__

18. Kyle is planning a senior prank. He is going to cover the football field with spray paint. He knows the area of a field is 121 meters^2. The sides of the field are the same. Kyle needs to know what the lengths of the sides of the field are so that he can make sure he has enough people on each side to get the field completely covered.   __11 meters__

19. Kristie is making cookies for a surprise party. There will be a lot of people there so Kristie needs to triple the recipe. The original recipe calls for ½ tsp. baking soda. How much baking soda does Kristie need now?   __1 ½ tsp.__

20. I am three-dimensional. I have 6 faces and 6 pairs of parallel lines. Each face has right angles. All of my faces are congruent with each other. Who am I?

   __cube__

www.ingramcontent.com/pod-product-compliance
Lightning Source LLC
Chambersburg PA
CBHW081620170526
45166CB00009B/3040